TRANSFORMING ANIMALS

TURNING INTO A FROG

by Tyler Gieseke

Cody Koala
An Imprint of Pop!
popbooksonline.com

abdobooks.com
Published by Pop!, a division of ABDO, PO Box 398166, Minneapolis, Minnesota 55439.

Printed in the United States of America, North Mankato, Minnesota

102021
012022

THIS BOOK CONTAINS RECYCLED MATERIALS

Cover Photo: Shutterstock Images
Interior Photos: Shutterstock Images, 1–20

Editor: Elizabeth Andrews
Series Designers: Laura Graphenteen, Victoria Bates

Library of Congress Control Number: 2021942432

Publisher's Cataloging-in-Publication Data
Names: Gieseke, Tyler, author.
Title: Turning into a frog / by Tyler Gieseke
Description: Minneapolis, Minnesota : Pop!, 2022 | Series: Transforming animals | Includes online resources and index.
Identifiers: ISBN 9781098241179 (lib. bdg.) | ISBN 9781098241872 (ebook)
Subjects: LCSH: Frogs--Juvenile literature. | Amphibians--Juvenile literature. | Animal life cycles--Juvenile literature. | Amphibians--Metamorphosis --Juvenile literature. | Animal Behavior--Juvenile literature.
Classification: DDC 597.8--dc23

Hello! My name is

Cody Koala

Pop open this book and you'll find QR codes like this one, loaded with information, so you can learn even more!

Scan this code* and others like it while you read, or visit the website below to make this book pop.

popbooksonline.com/turn-frog

*Scanning QR codes requires a web-enabled smart device with a QR code reader app and a camera.

Table of Contents

Chapter 1

Transforming Animals

A quiet pond sits in a forest. Frogs hop into the pond from the shore. Some of them make a ribbit sound! Frogs live all over the world.

Watch a video here!

Frogs are **transforming** animals. They cannot hop or ribbit as babies. First, they grow through four steps. The steps are egg, tadpole, froglet, and adult.

Life Cycle of a Frog

Every frog begins life as an egg. Female frogs lay the eggs in or near fresh water.

The American bullfrog can make 20,000 eggs at once. Only about 400 of those are likely to survive and **hatch**.

Many lay thousands of eggs at a time. The eggs are sticky and squishy.

Chapter 2

Tiny Tadpoles

Frog eggs **hatch** after one to three weeks. The babies are tadpoles. Each has a long tail, **gills**, and no legs. They live underwater.

Learn more here!

A tadpole eats small plants so it can grow. After about six weeks, back legs form. The tadpole can begin to hop.

The tadpole also starts to lose its gills and grow lungs. Lungs will allow it to breathe air!

Some adult frogs carry their newly hatched tadpoles to the water.

Chapter 3

Growing Up

At about twelve weeks old, the tadpole is a froglet. It has lungs and four legs. Still, it lives mostly in water. Froglets look like adult frogs in many ways. But they still have tails.

Learn more here!

Froglets use the **nutrients** in their tails as food. As they grow, their tails get

smaller and smaller. The tail becomes part of the body. This takes two or three weeks.

Frogs are amazing jumpers. Some can leap lengths 20 times their size!

Chapter 4

Colors and Croaks

When a froglet's tail is gone, it is an adult frog. Adult frogs breathe air and hop on land. Some have bright colors. These tell **predators** they are poisonous!

Complete an activity here!

Frogs live about six to eight years. An adult male will **croak** to attract a **mate**. The mates make new eggs. The life cycle starts again! Frogs are amazing **transforming** animals.

Making Connections

Text-to-Self

Which is your favorite step in the frog life cycle? Why?

Text-to-Text

What stories have you read about frogs? Did they include frogs at different steps in the life cycle? Why do you think they did or didn't?

Text-to-World

What is another transforming animal you know about? How is its life cycle similar to or different from a frog's?

Glossary

croak – to make a loud noise deep in the throat.

gills – body parts that help animals breathe water.

hatch – to break out of an egg.

mate – a partner animal of the same kind. Together they make new eggs or babies.

nutrient – a thing in food the body needs to grow.

predator – an animal that hunts other animals for food.

transform – to change into a new shape.

Index

Online Resources

popbooksonline.com

Thanks for reading this Cody Koala book!

Scan this code* and others like it in this book, or visit the website below to make this book pop!

popbooksonline.com/turn-frog

*Scanning QR codes requires a web-enabled smart device with a QR code reader app and a camera.